From sailboat to economical motorboat

La belle et le bouchon gras

Domi Montésinos

From the same author

ISBN 9798393683061

DEDICATED :

To all those who are not afraid towet their
shirts to make their ideas and dreams come true.

Table des matières

1-PREFACE

Living on the ocean and moving around is not particularly easy for a human being. Numerous clues quickly allow any individual tempted by the adventure to be persuaded.

As for me, being born in Saint-Pierre-et-Miquelon, France which is just south of Newfoundland, the sea is an exclusive and possessive lover, had hypnotized me, and bewitched me even before I knew how to walk.

Enrolled by force in her fascinating world, I never had any other choice than to learn how to live with her without being swallowed up by her. I had to constantly invent ways to prevent her from swallowing me up. This is probably the origin of this devouring passion for boats that I started in my early childhood, and which has never left me.

This compulsive nautical desire has led me to be hopelessly addicted to the aspects and facets of everything that touches the world of floating machines.

I have always been eager to understand how it is possible for a human being to manage to stay alive in this environment so unnatural for him.

The resources to be implemented and the technical solutions available are very numerous.

In a way that I hope to be pragmatic, this book aims to provide some insights and answers to the question that we have often heard lately:

"Why doesn't your new boat have a mast?"

Short question, short answer:

"Because I am no longer strong enough to operate a big sailingboat as it should be...

2- A BOAT TO LIVE IN

I made the choice to live on a boat a long time ago. I started dreaming about it when I was 10 years old.

It was at the Tarbes fair (south of France) a small Breton shipyard was exhibiting a 25 feet fiberglass boat.

This sailboat with an auxiliary engine was equipped with a cabin in which one could see a table, a bench and a camping stove... I leaned over to get a better look and my heart started to maximize his rpm:

- I could very well live in there!

I could already see myself there... Shortly there after, I declared to my mother:

"When I grow up, I want to be a sailor."

This child hood wish has remained engraved in me since that time.

However, I have spent long periods of time away from the sea and boats living on the land.

Mamilou and I owned a house in the south of France, for seven years. That's why I can say: living on land isn't really my thing.

There are two points in particular that bother me about this way of life.

First of all, this strong fact that homes have to remain fixed, firmly attached to the very place where they were erected. For the nomad that I am, this characteristic, which some people appreciate, gives me a feeling of imprisonment not very compatible with my thirst for freedom.

Secondly, as genetics has given me the great gift of being able to work with my hands, I'am rarely reluctant to improve this or that detail of our home. Alas, along with this good fortune, I have developed a certain aversion to the materials commonly used in construction.

Everything that is cement, brick, plaster, stones, honestly, "I don't like". It's silly, I know, but I'm not going to pretend to ignore it once I've clearly identified this weird characteristic that affects me.

As much as I enjoy working with wood, metals, resins and fibers, I'am quite put off by stony conglomerates.

Come on, a short riddle to lighten the atmosphere:

Among the following words, find theintruder: self-saw, welding machine, joiner, laminating roller, wheelbarrow...

In short, the habitat that suits me, being mobile and preferably floating, has always seems to be a strong welldesigned boat.

There are many ways to move your floating home. Two of them occupied me a few thousand hours: sailing and the thermal engine.

I won't mention the electric motorization, because I have only experience with electric motors on dingys.

Together with my wife Mamilou, we have achieved a seven- year journey around the world on a 64' sailing catamaran. We came back with the great satisfaction of having made this fantastic trip and the desire to continue to roam the seas. But as we got older a boat that required less physical strength was needed.

To sail a catamaran of more than 40' together requires a good physical condition as well as deep pockets...

A motorboat is much easier to handle with two people, and less expensive to operate.

We imagined it to be about fifteen meters long and propelled solely by machines, without mast or sails.

And so the "boat, sweet boat" that welcomes us now is a catamaran, without mast, driven by two solid diesel engines of good quality.

And we are very happy to be free of the enormous constraint of having a mast and it's rigging.

Which, let's not forget, is only useful when we are moving, that is to say ten to fifteen percent of the time, only (or even less...).

The skippers of the many sailing boats that are permanently travelling with their engines will not contradict me...

Explanation of the costs of sailing

The most common sailing system nowadays is made up of a multitude of components, each more expensive than the last.

First of all, the mast itself, the rigging, the sails and the numerous ropes that are used to manage it all. Without forgetting the elements the fittings that allow to master these maneuvers (winches, blocks, shackles...), and finally, the structural reinforcements, necessary for all this stuff to stay nicely in its place.

All this material represents, on average, equals something like a quarter of the value of the boat!

Let's take the example of a mid-range production catamaran that is about 8 years old and valued at 400 k$. Its lucky owners will have 90% of their time (the time during which they are at anchor or in themarina) to contemplate their hundred thousand dollar investment (the mast and its accessories) depreciating inexorably above their heads.

Why not, if they want to? But I think it's a high price to pay to be able to sayt his sentence, which I had made my own for years:

- I go on the sea by using sails!

And then, one more big advantage:

When suddenly, in the middle of the night, a raging squall occurs, as it is frequent in the tropics, the crew is sure to be awakened by the disturbing whistling of the wind in the rigging and halyards banging on the mast.

The sailing catamaran in our example is probably equipped with two moderate power diesel engines as standard.

After about ten years of sunny use, the set of sails and the complete rigging will have to be replaced.

The cost will be about three times the amount that would have been spent on carburant diesel and various consumables if we had sailed only with mechanical propulsion, over the same distance and at the same speed.

But I feel the anger rumbling in the skulls of intrepid lovers of beautiful sailboats, stunned by my boldness and disappointed to see me thus "denigrate"what I have practiced so much.

Don't get me wrong, I want to be clear. Farbeit from me to trample on what I used to adore what seems like just yesterday.

The pleasure of steering a sailboat by trimming it correctly and steering it with finesse and firmnesss always seems to me really irreplaceable.

My point is elsewhere.

What I say is that moving one's floating house is a different matter. And that's what this is all about.

As far as we are concerned, it is certain that moving a ship with machines is much simpler and less expensive than doing it with a mast and sails (at the same speed).

So, now we are planning to sail with the engines.

But then, what about the ecological motives that many lovers of beautiful hulls brandish as a pass to paradise? (This seems to me to be a hazardous objective. Because I am not far from thinking that the "Eden" is here and now, at least, for the luckiest among us and the most skillful at creating happiness by benevolently considering their glass like halffull).

Well, at the risk of shocking you again, I would like to share with you the idea that we do not pollute more (and may be even less...) than our "sailing with sails" counterparts!

This is on the condition that we travel at a speed lower than the speed of the hull of the boat (i.e., less than nine knots in the case of Lady't Bee).

Don't get angry too quickly, the explanations follow.

The key element of my demonstration lies in the production of the energy necessary for the comfort of daily life on board.

Obviously, certain choices had to be made as a result. The available surface on the roof of Lady't Bee was entirely devoted to the installation of a generously sized solarpark, which makes it possible to do without a generator.

This is no small thing.

A lot of people who like to run their genset for several hours a day for the production of their comfort energy, when it's not their propulsion machine that idles...

On a sailboat, it is difficult to cover the horizontal areas with photovoltaic panels. Indeed, a large part of the space is occupied by the fittings, and the rest is often shaded by the different spars, mast and boom in particular, and also by the sails themselves.

No gasoline on board Lady't Bee either, as the dinghy's motor is electric.

Of course, if we happen to sail in areas where the sun is usually shy, it could be wise to install two wind turbines.

This is not the case for us at the moment.

A few words about the storage of these "for free and easy" kilowatts:

There is no need to have a very big batterie's capacity. On the contrary, all you have to do is to take care, on a daily basis, to use the soft energies when they are present, that is, in the middle of theday.

Obviously, if you decide to use the water maker and washing machine at night, it won't work very well. But by running the big consumers between ten a.m.and four p.m., we can power the same equipment with this system as in a house of comparable size and standing.

There is nothing to prevent several devices from being in operation simultaneously.

The osmosis machine can produce sixty liters of water per hour, while the washing machine cleans the cloths. Also, if we have a friend visiting us, who likes espresso, the electric coffee maker and the tea kettle will be there. And, for cooking, the induction plates replace advantageously the gas, whose storage always presents a non-negligible potential risk.

As for the propulsion, a judicious choice of propellers, associated with the addition of Econokits1, were the key elements in transforming a slightly feisty "power cat" into a respectable "long range motor-yacht", which is capable of crossing the oceans in complete serenity.

Low consumption and considerable storage capacity have given Lady't Bee a wide range of autonomy.

Even with no long navigations, this feature is a great comfort and economy element, because it allows to provide carburant diesel where it is the cheapest...

---

[1] A commercially available French system that can be installed without difficulty to improve efficiency of a combustion engine, and reduce the pollution and consumption

3- OUR VISION OF THE IDEAL BOAT

In our minds, the multihull was the obvious choice.

This is for a number of reasons, the first of which is the initial stability, which is particularly appreciable at anchor.

Since we spend most of our lives in this mode, low roll sensitivity is a decisive factor.

And then, in terms of habitability, available space and comfort on board, the catamaran proved to be the best compromise.

However, I remain convinced that it is possible to do something very interesting with a trimaran base.

I excluded this way to respect the desire of Mamilou who does not like this option...

This "ideal boat" does not have a mast, nor any other aerial protuberance that could increase the dunnage2 to the detriment of the well-being and the marine qualities.

It draws the energy necessary for life on board from renewable sources: solar panels and wind turbines.

It has a trans-oceanic capacity.

_____

2 Frontal surface causing driving resistance

It is because I love more than anything the feeling of freedom that a good yacht inspires, able to go autonomously almost anywhere in this world.

The technical solution chosen for the propulsion is that of the diesel engines with shaft drive, because it is a perfectly known and mastered solution.

The two engines are installed in such a way that they can operate in turn in cruise mode, or both together if necessary.

This arrangement reduces fuel consumption and ensures that one of the engines is always available for inspection and maintenance.

The cruising speed of eight knots is obtained by turning only one of the two propellers at 65% of maximum engine speed.

In this configuration, the boat will still reach 12.5 knots, with the two motors at full throttle!

For reasons of cost, use and acquisition, we have decided not to exceed 15 meters (50 feet) in overall length.

And for reasons of comfort at sea and safety, we have set the low limit at 13 meters (44 feet).

4-HOW TO BECOME THE OWNER OF THIS WONDER?

Not easy.

There are very rare units on the second-hand market, and they are mainly large vessels.

They are "one-offs" with original features that often make them difficult to resell.

No manufacturer of production boats offers this type of yacht for the moment.

The closest thing we can find are catamarans, either sail or motor, whose architecture and design allow them to be transformed into high autonomy offshore vessels.

More or less considerable transformations are of course essential for this.

What does that mean?

To begin with, the removal of everything that is used to propel a sailboat or the removal of everything that is "anti-marine" in the case of a "power cat".

This last solution is what we chose.

We had previously studied carefully the case of some potentially compatible sailing catamarans, but finally we abandoned this way...

The "coup de coeur" did not tip the scales in their favor.

It must be said that the idea of buying a sailboat, for a considerable amount of money, and then to start by stripping it of a whole bunch of expensive elements, may seem far-fetched... Once you have removed the mast, the rigging, the fittings, the anti-drift fins, you find yourself the owner of a kind of "demolition yard"..., plus a bundle of junk on the quay that you will try to sell for a tenth of its value... This is the kind of project that requires a substantial motivation

On the other hand, technically, it is a very viable solution, provided, however, that the entire chain of propulsion equipment has to be seriously reviewed.

Exemplary reliability is obviously essential as we can no longer rely on sails to move around.

Starting from a "powercat", things are significantly different, even if the processes are strangely similar.

In this case, we will systematically have overly powerful engines, with their corollary: bulky, heavy, greedy, expensive to use.

Also, for many years, "terrace" cockpits have been flourishing on the second floor of many yachts in this category...

There is no doubt that this corresponds to the taste of a certain clientele, much fonder of idleness than of real sailing. These "boxes", ungainly and high perched, certainly do not favor the boat's marine qualities with this excessive weight in the tops.

Because of this last point, it took me many months of "reflection" before I was able to envisage our project on this kind of base.

5 THE LADY

In the end, we chose a "Moorings 474 PC".

This model of power catamaran was developed and manufactured in the Robertson & Caine shipyard in Capetown for the Moorings Company. They operated a large fleet at its Tortola's base, in the Virgin Islands. Then they sold the mafters six or seven years of charter.

Parenthesis: a large portion of their armada was destroyed in 2017 by Hurricane Irma. Many units were declared "wrecked".

I have spent a large part of my professional life building boats. At the beginning, it was with my own hands (and Malou's help).

Then, during another, much longer period, I managed people who wanted to do the same (but without Malou's help...).

So, because our budget was terribly tight, we chose this brave "LadyB".

After serving valiantly during its years of intensive charter, she had well deserved a solid "refit", implemented by a recognized professional.

Prior to its acquisition, I donned my expert's hat and listed the more or less obvious defects.

This list brought us decisive arguments, which Malou was able to exploit skillfully, to allow the purchase of this catamaran at its fair value.

At this point, all we had to do was roll up our sleeves and put in the 2,500 man-hours required for the face lift.

Feeling a bit weary from the weight of years, we started by asking for two or three quotes, in order to eventually share the joys and satisfactions of this "huge job" with a few courageous craftsmen.

The reception of the first quotations considerably modified our vision of the project.

It was quickly decided to go back to the good old method of our twenty years: to "do" a lot of the work ourselves...

Luckily, courage returned

Play on words

Then, we adopted the American flagged "LADY B", in Tortola, British Virgin Islands (BVI).

Being, by nature, not superstitious, I had no reason to think that changing the name of a boat can bring badluck, oreven happiness or anything else.

However, I honestly admit that I prefer to keep the old term... It's just the way it is!

It is not really a question of superstition, we agree...But, at the same time, I do not see why I would take risks when "you never know"...

In short, the time had come to separate us with the "stars and stripes" banner. It was for the benefit of this maple leaf which about of which its elixir marries so well with rum and lime that the battery syrup of Marie-Galante is quite jealous of it.

But the Canadian authorities told us that it was impossible to keep the name "Lady B", as it had already been given to another boat. What a pity!

It turns out that, during the whole period of negotiations prior to the adoption, Malou had taken to humming a famous song by the Beatles which sounds quite close to this name. (Let it be)

This is where the idea came from to get us out of this mess. "Lady B" became "Lady't Bee", which pleased everyone.

As soon as we were settled on board, a "to-dolist" was drawnup. It was certainly not a short list like "repaint the bathroom" or refill the fridge with a pack of beers...

There was a lot of "heavy duty jobs" in there... And, there was a lot of evidences to suggest that this was just the beginning.

To begin with, we had to go and tickle some sea spray as quickly as possible to find out a little more about this Lady's behavior "at sea".

A first navigation no fone hundred miles, right into the wind, in this case a brave tradewind of about twenty knots, informed us, without the situation, on the next few lines of the list.

Before we left Tortola, a decrepit old captain, all worn out from the carcass, moving with difficulty and a wooden cane, had detected our plan to acquire the Lady.

He told us in an aside:

- You're going to buy this??!!!!

But it's just a "Mickey Mouse" boat... It's only good for little cruises in the islands and luring naughty girls, but you can't go to sea in it..."

A philosopher, of course.

This was a bit of a hasty oversight of the fact that these boats, were designed in a renowned firm, and bear the signature of a well-known and respected architect.

Moreover, they all came to the Caribbean by crossing the Atlantic from their construction site in Capetown, South Africa. A small six-thousand miles "trip", all the same...

Rock'n'roll

This is the word that comes to mind to describe this Tortola/St. Martin delivery trip, although it was carried out in ordinary conditions, i.e. a trade wind of around twenty knots. It is that, perched on the second floor, where the cockpit is, the discomfort is obvious.

Impossible to take shelter from the rain in the squalls! And it was absolutely necessary to hold on tight all the time to avoid being thrown out of your helm seat.

I already had a strong aversion, before, for the ugly look of this incongruous deck saloon, reproaching it for considerably uglifying the boat. This walk of a few hours did not lead me to change my mind.

Honestly, I couldn't see myself doing night "watches" in this context.

It quickly became clear that the removal of this "wart" was necessary and as soon as possible.

This first navigation was the only one accomplished in this configuration.

A week later, I had dismantled all the instruments, electrical cables and hydraulic hoses, and moved the pilot house to the chart table, at the first floor, in the saloon.

The first "copious" project, from which some changes of the same kind were to follow...

6-AUTOPILOT AND PANORAMIC VISION

The operation "descent to the first floor of the navigation station" had made a "left for dead", the steering wheel.

The Lady could now only orient herself with the help of her two motors, or with the autopilot.

This situation was quite bearable, but a bit unusual and not very recommended.

It decided to install a second pilot, equipped with a knurled knob that controls the rudders by simple rotation. In short, a mini steering wheel of two centimeters in diameter...

To which I soon added a "tiller", a kind of "joystick" that directly controls the hydraulic cylinders that move the rudders

The new driving post thus formed soon proved to be much better adapted to our style of navigation.

We are safe in all weathers. The movements of the boat have a lesser amplitude, and everything is within reach (you can, simultaneously, steer the boat, sip a beer and enjoy a good meal prepared by Mamilou...)

Equipped with a well-made office chair, it welcomes you for hours without fatigue.

On the other hand, as far as visibility on the outside was concerned, there was room for improvement...

This led to a new "slight modification"...,but on a large scale: the windshield

The two original front windows were separated by a massive piece of polyester, an integral part of the deck (in terms of construction).

This deck was made from a common mold for the Leopard 46 (sailboat) and the 47 PC (power cat). This strong piece was designed to support the mast's foot in the sailing version.

However, for the motor-yacht, this appendage had no more interest.

This massive element was also in charge of keeping the two huge methacrylate panels in place with the required watertightness in such circumstances.

Alas, our brave Lady't Bee was far from being irreproachable on this point. To tell you the truth, she was leaking like a diaper - second hand panties...

Add to this already "annoying" inconvenience, the visual obstruction inevitably induced by the piece of laminate perfectly useless in the absence of a mast.

And here we have all the necessary ingredients for a firm and unquestionable position:

"We're going to fire all of this!"

This was followed by a waltz of chain saws, disk saws and some others cutting instruments...

This worried Mamilou a lot, besides making her angry because of the dust that was spreading "almost" everywhere.

"Don't you feel like you're cutting a bit much here?"

"It's the exact opposite... I'm filled with doubt. Have I even taken enough off?"

Surgery is not homeopathy. You have to "hit it" with vigor and determination, otherwise you'll never see the end of it.

In two times, three movements...

Well, OK, let's say, in "a hundred times, a thousand movements", a kind of panoramic wheelhouse appeared.

It made Lady't Bee ten years younger and gave her a host of fabulous advantages: thus, the glazing in "securit" glass, the electric windshield wiper, the roof against the greenhouse effect.

The latter supports six hundred watts of solar panels, and offers a large clear view to the front.

It's quite simple: every new visitor entering the cabin for the first time invariably utters this interjection: "Ouahhhh!

Storage spaces

Standard boats, whatever they are, rarely offer enough closet and storage space to sail a long distance, let alone live aboard.

On the other hand, in a 47' catamaran, the volumes exist for this. They are simply badly used if not "unexploited at all"; just neglected.

Here again, the landing of surplus equipment will turn out to be quite salutary. Especially when they are located in easy-to-reach places, as was the case with an air conditioning unit under the saloon seat.

One of the most "cost-effective" solutions for creating storage areas is to put up shelves that are well distributed.

Many of the original volumes are almost unusable because they are not partitioned.

Of course, they can be filled in anyway by stacking things, but later access is then inconvenient.

Installing shelves and cross partitions solves the problem well, but unfortunately at the cost of singular acrobatics, because it is particularly difficult to work in this kind of space.

The most prominent examples aboard Lady't Bee were under the galley countertop, as well as in the aft peaks and in one of the forward peaks.

One of them was dedicated to the accommodation of the skipper in the "charter" mode.

The rear spaces each received their own custom-made staircase equipped with sliding plastic basins under each step.

Thus, between the development of existing spaces and the methodical tracking of all "undesirables" we have created enough space to host, in turn, a whole bunch of junk... Maybe as futile as the previous one, but, at least, it is ours...

7 ON-BOARD ENERGY MANAGEMENT

The Moorings 474 PC is equipped with a nine-kilowatt diesel generator as standard.

Added to the two propulsion machines, this third engine, to be maintained and fed, did not attract my sympathy.

About the respect of the environment, I am not a narrow-minded environmentalist, but I desagree these consumers of fossil fuels.

This machine, belonging to a rather well rated category, had five times more operating hours than its colleagues assigned to navigation (15,000 hours).

There are two explanations for this.

Being operated in the BVI, these boats move little, as there are many anchorages in this group of islands, not very far from each others.

They receive American customers, who cannot live without air conditioning.

As a result, the generator was running night and day, except for the daily traveltime necessary to get from one harbor to the next, all this in order to supply permanently the three "air conditioners" of the board.

Well, we don't have any merit for removing two "aircond groups" because we don't like the air conditioned.

And then, the systems of air circulation which we installed are sufficiently effective not to need these very polluting cooling systems...

Once the upper floor was duly cleared of its plethoric and heavy "gadgets", a magnificent area of about twenty-five square meters was vacant, and therefore, able to accommodate a respectable solar park.

At this stade, we took a paper, a pencil, a calculator and an internet connection to inquire about the state of the art.

It was soon obvious that the solar production would be perfectly capable of supplanting the noisy machine to largely provide for the energy needs of the yacht.

Hence the new project:

Remove the genset and two of the three air conditioners.

Instead, thirteen photovoltaic panels of 300 watts each, a set of lead/acid batteries of 900 ampere-hours and 7 charge regulators were installed.

Two inverters of 3000 watts each serve a 220 volt electrical panel exactly the same as the one you would find in a house connected to the land grid.

Thus equipped, Lady't Bee has all the modern comforts: hot and cold watertaps, washing machine, coffee machine, kettle, computers, watermaker and all that sort of thing, plus an ice-maker.

A clever little device, from Victron, allowed me to connect a relay controlled by the voltage of the battery pack and its charge level.

So, if a cloudy sky limits the solar contribution, the supply of the waterheaters (or of the air conditioning) is "cut", putting us safe from the risk of approaching the night with an insufficient reserve!

As for the question of lighting, all the light sources on board have been changed to LEDs.

All lit up together, they add up to less than one amp. How cool is that?

"So what was the point of keeping an air conditioner?"

Good question and I thank you for asking it.

It is for the beauty of the gesture...

Thus, in hot weather, on a very sunny day, this device can operate on solar energy. During the hottest hours, "it is the sun that cools us"! Magic.

8-HULL AND PROPULSION SYSTEM

I really had made plenty of projects around the idea of an electric motorization!

I consulted specialists, read articles and books, scribbled simulations and filled in pages of calculations...

The "reasonable" people have broken my enthusiasm.

I ended up siding with the serious ones, the ones who "know their stuff", those who have their feet on the ground..., while mine are better on the seas.

At present, I devour, with greed and a touch of jealousy, all the informations concerning the exciting journey of "Energy Observer". This adventure, at the same time human, scientific and concretely ecological, offers me a strong subject of hope concerning nautical activities...

When Lady't Bee was built, she was equipped with two diesel engines of 150 horsepower each.

These were calculated to give her a maximum speed of 17 knots, with a modest range of only a few hundred miles...

Nothing to do with long distance cruising aspirations.

To cross an ocean, it seems reasonable to me to count on autonomy of three thousand, or even three thousand five hundred miles. Even supposing to adopt a speed much lower than the theoretical capacities of the machines, it is not possible to obtain a sufficient autonomy with such powerful engines.

It is important to understand that you absolutely cannot sail permanently with "idling" diesels (6.5 knots, idling, originaly…), or you will damage them3.

Fortunately, by acting on the different parameters involved, many things become possible that will allow us to keep these wonderful Cummins built to last more than thirty thousand hours.

In all likelihood, and barring a major accident or great misfortune, they should logically outlive us, which makes me very happy. But before everything, I need to mention somethinh very important.

The fuel consumption of a motorboat is inversely proportional to its waterline length and directly proportional to its weight.

Conclusion:

Let's start with a good weight loss program. It is not the most expensive thing to do, and the effectiveness is tangible and immediate.

The displacement

Everything that has no use on board must go...

The limit is constituted by the consent or disapproval of the co-owner. If this person is your wife…take care…

---

[3] At low engine speeds, carburation is incomplete and unburned particles accumulate in the exhaust system until they clog it.

For Lady't Bee, it was more than two tons of "trinkets" who were thus dismissed.

Some significant examples:

Toilet facilities…

No less than five toilet bowls, four of which are electric.

When you live with two people, that's a lot of spare parts.

We should have established a schedule for the use of the different "vases"... In order to be sure that they are all used on a regular basis, otherwise, the one who would not have received enough visits would be sure to break down.

It's too much! Constipation elevated to the rank of major scourge of the cruise! Classy.

Anyway, after consultation, we fired two of them (I was leaning towards three, out of pure selfishness, because I'm the one doing the maintenance, but...)

The *bazar* on the second floor

The space was very pleasantly arranged.

In fact, many boat show vendors are winning sales with this incongruous appendage and, above all, with its lounging facilities.

The problem was that this joke weighed almost a ton!

A load whose center of gravity was located at more than fourteen feet high!

In short, all of this joke represented almost half of the overweight.

Another thing:

As she was not originally equipped with a watermaker, Lady't Bee had four tanks inside her, allowing her to carry 1.2 tons of water!

Thanks to the installation of an osmosis unit capable of supplying 15 gallons per hour, the storage capacity has been reduced to 100 gallons.

The list is long of the disembarked elements that Mamilou sold all at small prices.

Floatation length

As for the waterline length, we have added sixty centimeters to the back of the hulls, in the form of two nice skirts.

They are, moreover, welcoming docks for the dinghy. Remember that this element is a factor in favor of speed and thus helps to minimize consumption.

The propellers

Once these small works were done, as a preamble, we had to attack a very serious subject, the propellers.

Big topic, slightly worrying.

The propeller suppliers, some of whom are obviously very competent, sometimes show a rigor that borders on rigidity.

These people have the habit of using calculation formulas, known by all, to define the diameter and the pitch of the thrusters they propose to their customers.

In the vast majority of cases, they have one and only concern, relayed by the engine manufacturer concerned: to move the boat forward as quickly as possible by making the best use of all the horsepower available "at the shaft".

Far-west atmosphere. Lucky Luke and Jolly Jumper in propeller's blades merchant costumes... I'm kidding, of course.

Compared to my problem, which is to go slowly, without damaging these machines that are much too powerful for the seven and a half knots that constitute my "target speed", it was not goingwell.

A real dialogue of deaf being established for a good moment with my interlocutor's calculators, I took the decision to define myself all alone "that is what I must do".

The Internet came to my rescue.

I had to read a few books on the subject.

After having imagined many figures and as many curves in my head, I fixed the characteristics of our future propellers.

Then, I ordered two units of a folding propeller model with the diameter and the pitch that I had determined.

As the selected supplier forced me to sign a release beforehand, my order was finally taken into account.

The pieces came to us, and I mounted them.

Good news amigos!

Everything went as planned. The results were present, proving that the people who wrote the guidelinest hat I used to decide are reliable people. I would like to thank them for that.

Attention: I specified propellers with "folding blades"

This point is essential.

Using very modest powers (about 23 horsepower to move forward at seven knots, without wind and without current, it is essential to be able to do it with one engine. Otherwise it means, in the case of Lady't Bee, that we can imagine two 150 HP engines working at about 25 horsepower... (Hardly more than idle!) It is therefore imperative to eliminate the drag of the second propeller when it is not in use.

For this purpose, folding blades are much more suitable than their feathering competitors, because the efficiency of the blades is poor due to their flat shape. This is very bad for the consumption. An essential component in determining the propellers was to significantly reduce their pitch.

Thanks to the advice of an excellent motorist, a Cummins agent, I started with the idea of calculating the pitch by setting a minimum engine speed below which it seemed desirable not to go.This rpm led to a speed that I defined as my "cruising"target.

Then, with the engines "at full throttle", all the available horsepower is not used. And as a result, the boat's maximum speed is far below what it could be if all the power possible was used with propellers designed for it.

But, it suits us very well! Since what interests us is to consume little in order to have a big autonomy and range.

Some significant figures to set the ideas: cruising speed 8 knots at 1800 rpm, and maximum 12.5 knots at 2800 rpm.

For a sailor with sails, this is "unheard of" (except with competition boats).

Groundplates

Originally, Lady't Bee was equipped with amost hideous and unsavory grounding devices. It was the propeller's chairs that received all the electrical "grounds"!

I didn't like this dangerous connection at all.

So I quickly replaced it with an efficient and economical system.

Flat anodes have been mounted on the planking on each side. Now, they host these nice yellow and green wires having all the attributes of the rastaflag and it's very nice...

The rudders

Once this was settled, there was still one important modification to be made: to enlarge the rudders.

Indeed, when sailing with only one engine, the rudders are necessarily stabilized at an average angle of incidence that compensates for the asymmetry of the propulsion. In otherwords, since we are pushing on one side only, we are going in the wrong direction.

The original rudders were sized to act at about 15 knots, so they are a bit small for the seven to eight knots we are interested in.

On Lady't Bee, I have increased them by thirty percent. I think that an increase of fifty to seventy per cent would have been even better (it can still be modified).

In the current configuration, the average angle of incidence at a speed of eight knots is about four degrees. Not too bad...

The Econokits

I cannot close this chapter without mentioning the considerable importance of the pollution and consumption abatement system named "Econokit".

I already used it for years, with satisfaction, on board our previous catamaran. As it was a sailing boat, the impact was less, but any savings in fuel and contamination are good to take.

The principle consists in introducing, in the intake manifold, a mixture of air saturated with moisture having undergone a molecular transformation by passing through a reactor.

The "reactor" is heated to three hundred degrees.

The objective is to obtain a more complete combustion with a reduction of unburned material.

The profit made in terms of consumption is usually twenty-five percent on machines of "classic" design, and can exceed forty percent on slightly older machines.

It is probably different on modern engines with electronically controlled injection, of the "commonrail" type. But I have not experienced this.

On board Catafjord, a 65'sailing catamaran, I had achieved forty percent, with the 45 horsepower diesels.

With Lady't Bee, I had a harder time doing this, as the engines are bigger, more expensive and equipped with turbos.

After extensive testing and modifications during a 3000 miles trip from St.Martin to Cuba, I achieved over thirty per cent fuel economy.

Then, after optimization, we went down to a consumption of 0.6 liter/mile at an average speed of 7.5 knots (about 1 gallon/hour).

These systems are often used on tractors and other agricultural machines. In these cases, it is easier to heat the reactor. It is sufficient to simply attach it to the exhaust pipe with metal clamps, as diesel engines release exhaust gases at a temperature of around 450°C.

However, it is not possible to do this with marine engines, as their design is governed by different rules, and there is no accessible place where such a temperature is available.

So I had to heat my reactors with electricity.

I had to make a custom heater myself using tungsten wire recovered from a toaster that I had boned. My numerous tests finally led to satisfactory results.

Then, once the experiment was completed, I had specific resistors made by a manufacturer in order to implement a more professional and secure installation.

Exhaust gas expulsed underwater

Another small point that might seem like a "detail", but is not: the exhaust outlets.

Originally, the burnt gases are led to the back of the hulls, on the outside, and come out halfway up the planking through beautiful stainless steel pieces.

At the boat show, it's superb!

In everyday reality, it's always dirty...The fumes deposit a black halo all around, and the metal drools a rusty ooze that doesn't look good on a unit as pretty as Lady't Bee.

An interesting solution we have adopted is to replace these parts with underwater exhausts.

It is necessary to take care to arrange them a little behind the propeller and shifted laterally towards the outside.

The dimensioning of the tubes, in terms of thickness, must be well realized.

They need to be comparable to the planking, because their solidity must be irreproachable, and they must come out about twenty centimeters below the waterline.

Add to this a conical deflector placed just in front of the gas outlet which will suck them by depression in navigation. It is, in fact, very important not to generate back pressure in the exhaust pipe.

All of this, properly realised, confers three considerable advantages:

Exhaust noise is reduced

The gases are evacuated in the wake and go out far behind the boat, thus never coming back to bother the crew.

The planking always stays clean.

However, since I no longer have visual control over the cooling water discharge, I had to install alarm sensors that detect a possible increase in temperature in the bubbler pots.

Water and fuel capacities

Originally, the Leopard47PC is equipped with two diesel tanks with a capacity of 150 gallons each.

And, for water, four tanks total 350 gallons.

After the addition of the water maker, storing such a large amount of liquid was no longer necessary.

So we kept a 100 gallons tank for fresh water and unloaded another 75 gallons tank located under the bunk that was removed (to become an office). Then we transformed the two remaining tanks into fuel capacities, i.e.175 additional gallons.

These modifications have increased the range to 2300 miles.

The service battery bank

A detestable solution, very common among production boat builders, is to build up service battery banks by connecting 12-volts cells in parallel.

It's easy to do and the batteries can be replaced with just about anything, available in all the shops aimed at motorists and other consumers.

Unfortunately, in doing so, you have to check out very frequently, as this is a provision that destroys batteries very quickly (three to four years).

This is for the simple reason that the different elements are never perfectly identical, so the best performing units discharge into the worst ones.

The only truly professional solution is to purchase elements of the chosen capacity and lower voltage, and then assemble them in series.

For example, if you want to equip your boat with a 12volts/700Ah, you will be able to opt for 3 batteries of 4volts/700Ah each, connected in series.

Or, 6 cells of 2volts/700Ah, always connected "in series".

Such an arrangement made of good quality products (as often found in solar dealers) will have a life span of about ten years.

On the other hand, a solution consisting of three 12volts/220Ah batteries connected in parallel (i.e.12volt/660Ah) will not last more than three years.

And let's not forget to point out for those lucky enough to sail on catamarans, that these boats, which are not very prone to acid leakage, have no interest in doing without the good old lead/acid batteries.

They are a heavier than modern "lithium" ones, but much cheaper. If it is to "save" 150 kgs, there are many other ways to do it for free...

9-OUTDOOR FACILITIES

The bar of the staircase

The installation of the solar park, on the top of the roof, had made obsolete the pretty spiral staircase which gave access to it, originally.

However, in spite of its undeniable charm, this "thing" having no more use, mobilized a considerable volume within our privileged living space.

So I had conceived by myself, quietly inside my head, the project of dismantling it with a chainsaw, in order to implant, why not, a nice bar!

This was done in order to receive with dignity some of my friends who are sensitive to this kind of attention.

One day, when I thought it was the right time to tackle this essential task, I decided to tell Mamilou a few details. I really thought that at some point she should have known about it, and that it was perhaps not wise to wait until she had the result in front of her eyes to take her opinion...

Ouhhlàlà, houlala...

I had to face a severe veto. A very negative "no".

I came up against a granite wall, the same one from which the spiral staircases leading to the pulpits of the Breton churches are made.

Slightly knocked out by the vehemence of her attitude towards such an obvious project, I sulked for a few seconds, just enough time to regain my composure. Then I switched the productive area of my brain to "hurry up and invent something clever before it turns into an argument". Fortunately, the sky came to my aid (under the impulse of Bacchus, perhaps...). And I was able, a few days later, to come back to my beloved with the following proposal, which I had taken care to present with humour.

The idea was to leave the two top steps intact, as well as the front of the very first one at the bottom.

The three others would be the object, not of a guilty destruction completely unacceptable (how did I not realize it from the beginning?), but, on the contrary, of a simple, provisional "deposit".

During this one, they would have to undergo the ablation of a few square decimeters... Then they would be assembled together in a delicious artistic posture.

What would not fail to confer them a particular aptitude to become what they would never have dared to hope, even in their most delirious dreams: a front of bar for Lady't Bee.

Ingenious, right? The idea appealed.

And a few weeks later, the refrigerator welcomed its first beers.

The icemaker was spitting out its first frozen cubes and the "Bar de l'Escalier" received its first guests.

Now that this major element of Lady't Bee is completed and well integrated, I thank Mamilou for saving me from committing some hideous installation in place of this beautiful bar.

The cockpit

In general, onboard a cruising catamaran, the cockpit is the most frequented living area.

On Lady't Bee, the installation of the solar park had already significantly increased the comfort of this space, which had become insensitive to rainfall since the staircase was condemned.

The installation of translucent side panels completed the transformation of this area into a living space that can be used in almost any weather, at anchor or at sea.

Only a heavy rain coming from behind can deprive the crew of a part of this place's quietness.

We have thus almost doubled the surface of the "square" by adding this"veranda".

With just a handful of dollars and a few hours of labor, the result was spectacular.

However, this time again, as light inconvenience added a few lines to the "list of essential works"...

We were running out of fresh air.

The heat climbed in our super "patio", exactly as the sun rose in the sky, until it became the hottest place on the boat.

The veranda was becoming a greenhouse! And this was tightening my heart.

I resolved, to remedy this, to capture the coolness that circulates permanently under the cockpit.

This one has two origins: first, it is in the shade, and then the flow accelerates while passing under the front platform what lowers the felt temperature.

Lady't Bee has two large trunks, opening onto the deck, just forward of the front window.

These volumes offer a front face conducive to the mounting of two small opening hatches, recovered during the remodeling of the front roof.

Thus, even by heavy rain, it enters, through these orifices lots of fresh air, but never water.

This air is then distributed throughout the boat through judiciously placed vents.

The fresh air supply in the saloon is therefore quite substantial.

Unfortunately, it is still insufficient to cool the cockpit area.

Of course, these "gills" are hermetically sealed before each cruising.

I completed the device by making a plastic tube "snorkel", with an erectile part that comes out under the floor when anchored and that is retracted before going sailing.

Last "ventilatory" refinements, two side scoops, with aeronautical looks, located at the level of the roof, complete this device which gives, at present, great satisfaction.

The davits

The ones that originally equipped the Lady were a remarkable concentration of defects.

It seemed as if the case had been entrusted to an individual with a special aptitude for this kind of exercise, charged with designing a striking example of all possible mediocrities on a single object.

And the guy had succeeded magnificently! A master, for sure. Everything was there: it was ugly, heavy, inconvenient to use, and certainly difficult and expensive to manufacture... A synthesis!

I was very satisfied the day I delicately put all these parts in the garbage can of the construction site.

Luck smiled upon us once again in the accomplishment of this work by putting on our path a very efficient handworker with whom we got along well, Stéphane.

We made an effective team together.

He is a slightly gruff guy, full of humor, common sense and expertise in working with stainless steel. We made a sober and practical unit that allows the dinghy to be raised and stored, while serving as a support for a set of four solar panels, which extends the shelter of the cockpit roof.

A great success and an interesting example of function integration.

Strong, esthetic and light.

The mast

Lady't Bee, the flirtatious, has put a feather in her hat!

A nice mast came to overcome his cap in a triple purpose:

Carry the anchor light, fly the flags and silence the talkers who tell us:

"And then, where is your mast?"

This last one is removable in order to allow Lady't Bee to pass easily under many bridges...

Air draft:

3.7 meters mastdown

4.9 meters erected mast

A validation cruise

All the modifications planned by the executive committee had not been yet encompleted... At this stage, there were still several weeks of work to complete the transformation of "Mickey's boat"into a true transoceanicvessel.

However, all the major points had been addressed, at failing that being perfectly completed...

Then, the season being propitious for a good cruise in the Caribbean, the construction was put on hold.

It was time to offer a few months cruise in the northwest islands...With the visit of Cuba as main objective.

We were right to do so, because on the way back, after many days of sailing, spent sailing up a strong trade wind, it appeared necessary to equip Lady't Bee with a system to seriously limit the pitching.

Thus, after overdozens of hours of progress in a posts planting mouvement, I came to design bow bulbous specific for this application.

It's incredible how the brain can works at its ease during those endless hours of wakefullness spent scrutinizing the horizon.

In the absence of any external disturbance "polluting the mind", our senses register a large quantity of "raw" datas. A few hours of sleep on this, to let the processor "grind" in peace, and solutions appeared as if by magic!

The anti-pitch bow bulbous

Thus, at the following year's refit, the slender and agile Lady't Bee found herself equipped with a pair of underwater appendages that reduced her pitch by about 40%!!!

Built on the technical area of Carriacou, they are made of marine plywood / PVC foam / fiberglass (occasionally reinforced with carbon), all assembled with epoxy resin.

The principle of operation is simple:

Their longitudinal sections are in "Naca" type profile, to ensure a minimum resistance to advancement.

Their cross sections are ovoid for the upper half and flat below.

The lower part has a slight spatula (like a surfboard) and is "anguled" of 4° to the front.

Thus, during pitching movements, the underside opposes the sinking. On the otherhand, the upward movements are little slowed down by the rounded shapes of the top.

These "appendixes" can only operate properly if they are permanently immersed.

Their rear parts are fixed directly to the hulls, while as carbon fiber reinforced legs holds them in the front part.

The result is enchanting!

Conclusion

Here is a brief description of the various adjustments that have made it possible to transform an ordinary "powercat" into a true offshore cruiser.

Certainly, these tasks are not within the competence of a simple handyman...

However, the skills to carry out such a "refit" exist on the market, even if they are not found every where.

The result is areal adventurer's boat, capable of taking its crew to the open sea, for distant cruises.

And this is in conditions of comfort and safety quite uncommon in units of this size.

Before and After

The original heavy davits

Modification of the roof: big job

OUAHHH

The original generator, replaced by the solar park

Propeller and rudder protection

Econokit and its 12 volt electric heater

The staircase bar with its windsock connected

Methacrylate side glazing

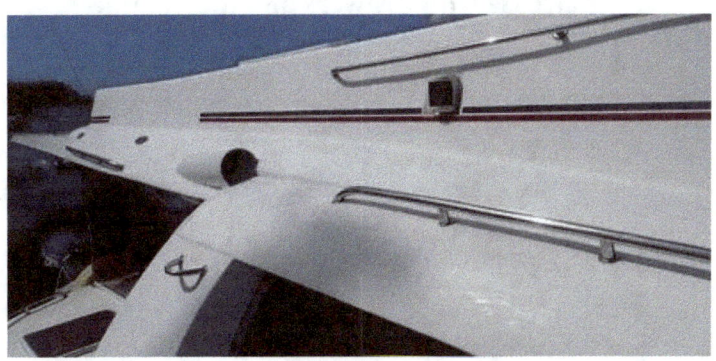

Cockpit side air intake

Staircase with drawers and diesel tank

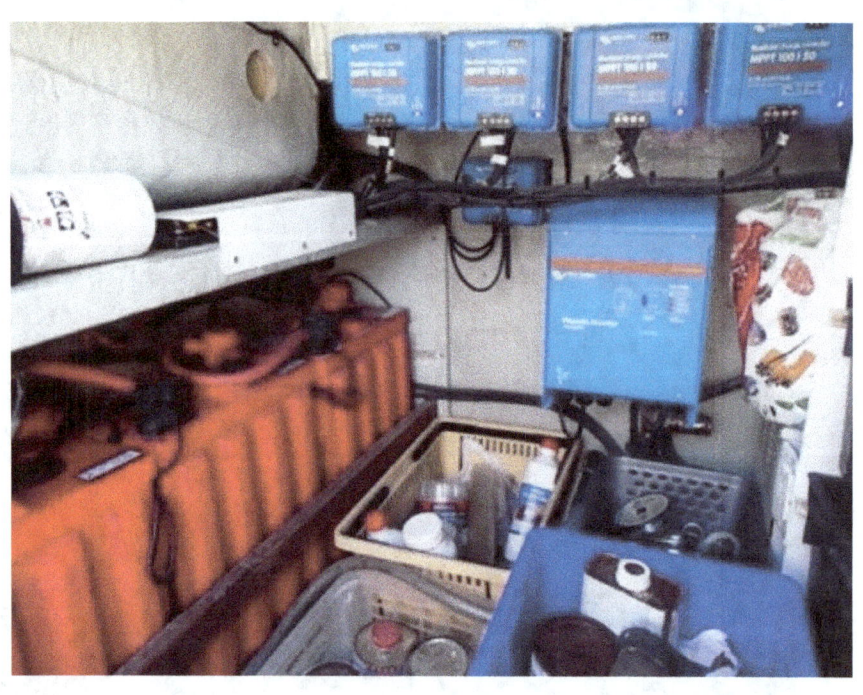

Battery bank, solar regulators, inverter

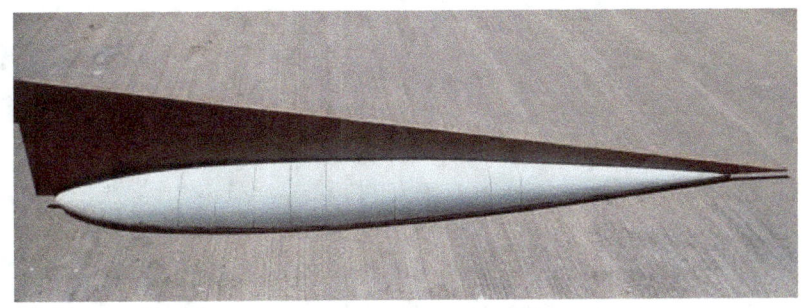

Structure of a bulb before stratification

Anti-pitch bulb

Underwater gas exhaust

The kitchen

The office and music room

The outdoor lounge

11-ABOUT THE AUTHOR

Who am I and where did I come from?

Born on October 5, 1953 in Saint-Pierre-et-Miquelon, my name is Dominique Montésinos.

Of my four grandparents, only my paternal grandfather, who be queathed me this beautiful Iberian-sounding surname, was not from Saint-Pierre. How can we be surprised by my great passion for boat?

Very young, I wanted to be a sailor.

Some health problems prevented me from attending the hydrography school in Nantes, but I still sailed for two years on various cargo ships as a mecanic student officer.

As a teenager, I was interested in electronics. After the merchant navy, I went further in this direction by following the courses of the Electrical Engineering section of the IUT of Lannion. These two school years did not bring me a job... But much more: a wife, who has accompanied me in (almost) all my peregrinations since 1973!

Having quickly established the axiom that we needed a good boat without delay, I became a builder out of necessity.

Amateur, first, with the complete realization of a 35 feet steel monohull, then, soon professional, in the work of aluminum, and then in the implementation of composite materials.

The demon of competition has guided our lives for five years. We frantically went from building to sailing.

The next "slice of life" was much more "standard". With the creation of our own companies, we had all the leisure to use our energy and our capacities in profusion... Alas without enriching us pecuniarily, as it was the goal...

Then, our youthful project was reactivated. It was the same one that had pushed us to build the steelboat of our first years.

The "transats", took over, slyly. So, as we approached 50, we reactivated it.

At the age of 52, I ended my career at Jeanneau to sail around the world with Malou.

Now that's done! Time to retire.

I decided to become a writer after the success of my chronicles during our seven-year trip around the world on a sailing catamaran.

And then I discovered that I liked itand, above all, that I enjoy being read.

ACKNOWLEDGEMENTS

Thank you to the many people who have played a role int his adventure. As well a stoall those who have, in one way or another, participated in the elaboration of this work.

I give a special thanks to **Malou** whose help and support are always so precious, and who took most of the pictures.

I also dedicate a moving and grateful thought to certain "bouchon gras"[4] and other officers of the Merchant Navy who taught me so much. It was so long ago that I still remember them...

Off course, I can't forget my very helpful friend "Clif", the best American sailor man I know, who participated in the translation from French language.

---

[4] Bouchon gras: nickname, synonymous with mechanic, given to sailors who work at the machine

www.ingramcontent.com/pod-product-compliance
Lightning Source LLC
Chambersburg PA
CBHW070457220526
45466CB00004B/1866